STEM ICONS

Celebrating LGBTQ+ History

Written By

Moni Singh

Creator of STEM for Kids

ISBN: 979-8-9878360-3-3

To my children Arnaav, Mansi and Ansh,
my papa and my mummy watching from her
heavenly abode.

Improving Children's Health

Sara Josephine Baker

US Federal Children's Bureau Founder
Josephine Baker was a public health pioneer.

A feminist, lesbian, suffragist and Doctor Jo, as she was known,
initiated programs for disease prevention and education.

When her father and brother died of typhoid
to care for her mom and sister, medicine will be her career, she decided.

Baker started as a physician in New York City as a private practitioner.
Then, she passed the civil service exam to become a medical inspector.

To succeed in the male-dominated public health administration of early 20th century,
she minimized her femininity by wearing masculine-tailored suits primarily.

In Hell's Kitchen, Baker took the opportunity
to help lower the rate of infant mortality.

1

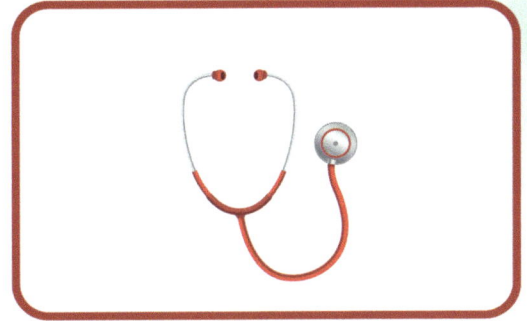

This New York slum had 1500 infant deaths weekly.
Most caused by intestinal infection called dysentery.

With a group of nurses, Baker got mothers into training
on how to care for their babies including bathing, feeding and
cleaning.

Commercial milk at that time was often contaminated.
Baker encouraged breastfeeding and provided milk that was
pasteurized.

Baker designed single dose eye drop containers to safely
deliver.
Decreased infant blindness from 300 babies per year to 3 per
year.

Babies were often delivered by midwives with no formal
training.
To ensure quality and expertise, she initiated midwife licensing.

She also ensured doctors and nurses at school locations
so school-age children could be routinely checked for
infestations.

Through her initiatives, Baker brought a revolution
for safety, health and hygiene for children.

Advancing Marine Sciences
Ruth Gates

Ruth Gates, born in Cyprus, was a coral research pioneer and marine biologist.
Her father worked in British military intelligence and mother was a physical therapist.

With her parents traveling constantly, Ruth grew up in a boarding school in an English city.
Earned her Bachelor of Science Degree and PhD from Newcastle University.

She was fascinated to learn that corals are animals that attach to the floor of an ocean.
Many rely on their relationship with plant-like algae, zooxanthellae, for energy and nutrition.

Zooxanthellae are photosynthetic, and use sunlight to generate energy. They live inside the coral tissue and give the coral its color in synergy.

Such corals grow in clear, shallow water, typically at depths under 60 meters.
Coral bleaching causes corals to become white due to stressors.

Stressors like climate change induced temperature rise cause the algae to produce toxins and coral to expel them before it dies.

Gates began studying corals in Jamaica, just as Caribbean corals were starting to die.
Continued her research at the University of California and the University of Hawaii.

Corals provide physical and ecological support to 33 percent of all marine life.
By 2050, only 10% of the world's coral reefs are projected to survive.

In response to bleaching events, while some corals couldn't live, Gates noted others seemed genetically predisposed to survive.

She reinforced genetic traits to breed "super-corals" through selective breeding
and transplanted them onto damaged reefs to withstand climate changing.

Elected to be the president of the International Coral Reef Society as the first LGBTQ+ woman, she advocated for inclusion and diversity.

Gates, now survived by her wife, is remembered not only as a brilliant scientist
but also as a mentor, public speaker, science communicator and an optimist.

Pioneering Tuberculosis Detection

Alan Hart

Born in Kansas and Masters in Radiology, Alan Hart
attended University of Oregon Medical School in 1917 to start.

He was respected by his community and peers,
until they found out he was transgendered, making him lose
careers.

Hart worked with Tuberculosis patients in Idaho and
Washington State.
Then, he moved to Connecticut in nineteen forty-eight.

Tuberculosis is an infectious disease.
It first attacks a patient's lungs, with ease.

Hart documented how it then spread via the circulatory
system,
causing lesions on the kidneys, spine, and brain in its victim.

Tuberculosis bacteria spreads among people not by gene.
But by coughing and sneezing in close proximity traveling in air
unseen.

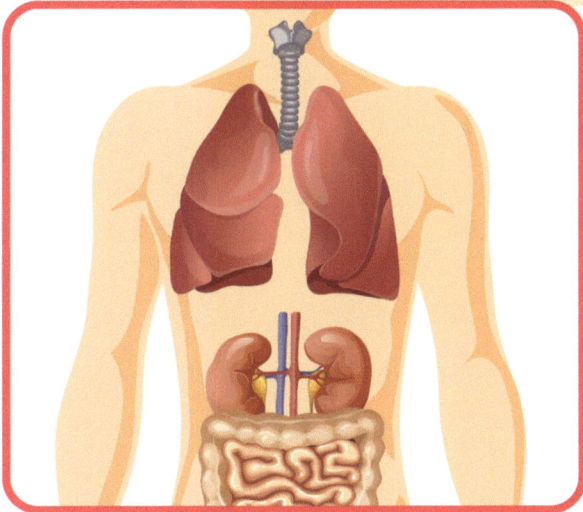

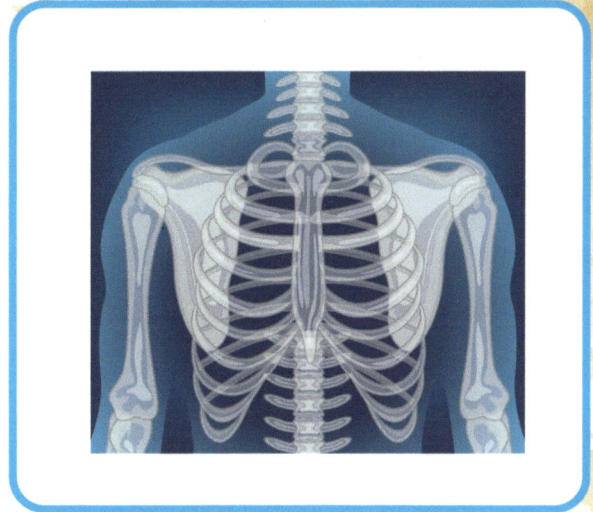

In the early 20th century America, TB was the deadliest infection.
Any hope for treatment required early detection.

Hart was the first person to propose
using x-rays to detect early signs before symptoms arose.

Used for fractures and tumors, X-rays were discovered when Hart was only five.
Thanks to Hart, X-rays could now detect TB, saving a patient's life.

Infected could be identified and their isolation
greatly reduced the spread of the disease in the population.

Hart wrote in medical journals and publications,
giving advice on TB's cure, detection and prevention.

Alan Hart established fixed and mobile TB screening clinics
and greatly improved recovery rates before the introduction of antibiotics.

Designing Electronic Circuits

Lynn Conway

Born in New York, physics graduate from MIT, Lynn Conway studied Electrical Engineering at Columbia University.

Her pioneering work at Xerox PARC in microelectronics led to worldwide chip design innovations.

A chip designer designs a circuit and then a mask set is created which is the master template of the design.

In semiconductor fabrication, the circuit design from the masks is transferred on a silicon wafer through a series of steps.

The fabrication process was very expensive. It required numerous masks each costing millions for a process so extensive.

So, designing chips for small schools and organization was very cost prohibitive due to lack of scale in production.

Conway founded the "multi project wafers" to solve the problem of cost by combining multiple small designs into one mask set.

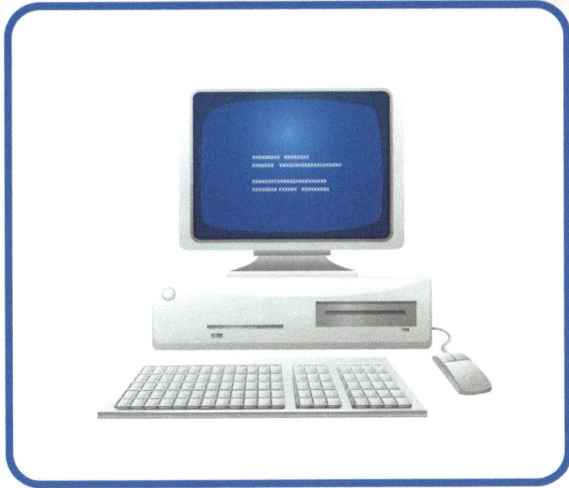

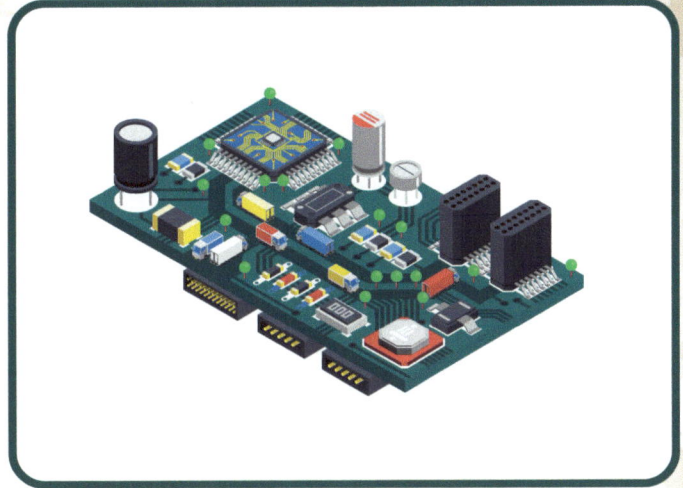

At IBM, her invention, Dynamic Instruction Scheduling,
led to improvements in computer architecture so compelling.

In the 1970s, Conway partnered with CalTech's Carver Mead
to write "Introduction to VLSI Systems", a landmark text indeed.

The book became the foundational text for chip designers in 120
universities by 1983,
leading to development of the microelectronics industry.

Behind her story of success were struggles due to her gender
transition.
She lost the only life she knew - her friends, relatives, job and
family - due to the decision.

At age 30, it wasn't easy to start all over again in a covert new identity.
Against all odds, she rebuilt herself as a symbol of human
spirit's invincibility.

Inducted to the National Academy of Engineering, Lynn Conway
holds the highest professional honor for engineers, they say.

Detecting Gravitational Waves

Nergis Mavalvala

Astrophysicist, Nergis Mavalvala was raised in Karachi, Pakistan,
becoming prominent
as a role model for aspiring female scientists in the Indian subcontinent.

Majored in physics and astronomy from Massachusetts's liberal
Wellesley College.
She researched condensed-matter and solid-state physics as she
acquired knowledge.

Even though she describes herself as a queer person of color,
the interactions she experienced were not hard for her.

For her doctoral research at MIT, she spent her days
developing a prototype laser interferometer for detecting gravitational
waves.

Interferometer, an investigative tool, takes light from two or more
sources to merge
And creates an interference pattern to measure, analyze and judge.

Created by massive accelerating objects, gravitational waves
are ripples in the fabric of spacetime traveling at the speed of light
rays.

Spacetime is a four-dimensional mathematical model that combines
into a single set, the three dimensions of space with time.

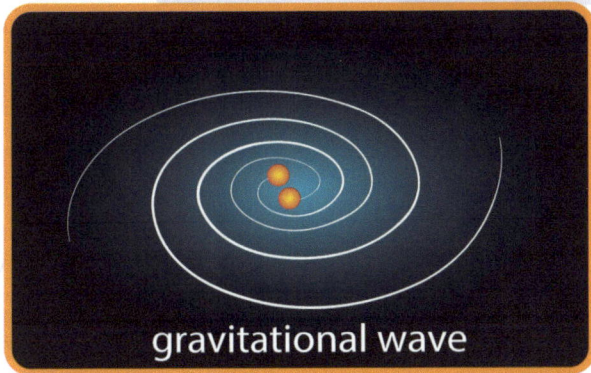
gravitational wave

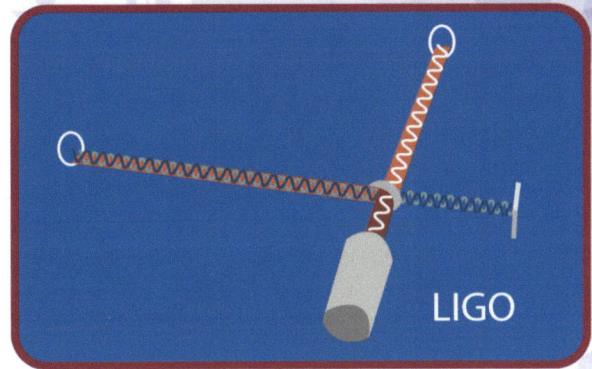
LIGO

Einstein's theory of relativity introduced spacetime distinct from ordinary space,
where a position is specified by three numbers, dimensions along x, y and z axes.

Mavalvala joined the Laser Interferometer Gravitational-Wave Observatory (LIGO),
an international project started by scientists at MIT and Caltech... ready, set, go!

LIGO included instruments 4 kilometers long, one in Hanford, Washington
and the other in Louisiana in the town of Livingston.

LIGO directly detected the gravitational waves in 2015 generated by two colliding black holes ... sight unseen!

Gravitational waves can give us a glimpse of neutron stars collision and other universe's major events like the Big Bang and a supernova explosion.

LIGO's discovery is one of the greatest scientific achievements... no gimmicks!
The work Mavalvala contributed to was awarded the 2017 Nobel Prize in Physics.

Researching Brain Function

Ben Barres

Ben Barres, ex-Chair of neurobiology at Stanford, got a BS from MIT,
an MD from Dartmouth and became a distinguished neuroscientist.

Though his interest for science developed in school,
he persevered because as a girl then taking science courses wasn't
cool.

During his residency, he observed a connection between
neural degeneration and patterns of brain cells making him keen.

Neural degeneration as in Alzheimer's is the loss of nerve cells
impacting
cognitive abilities like memory and decision making.

While there are 9 billion neurons or nerve cells,
nine out of every 10 brain cells aren't neurons but glial cells!

Barres decided for PhD in neurobiology at Harvard Medical School,
Boston
to research the glial cells or glia; their structure, function and
distribution.

While previously considered simply as the brain's biological glue,
Barres showed the glia played an important role in brain function too.

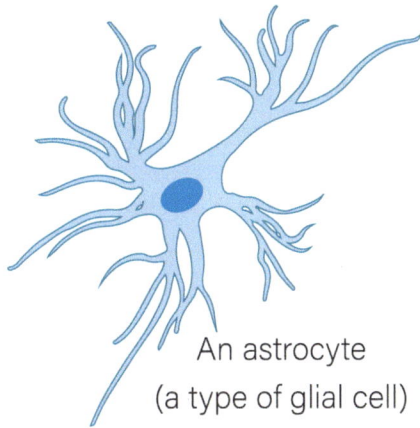
An astrocyte
(a type of glial cell)

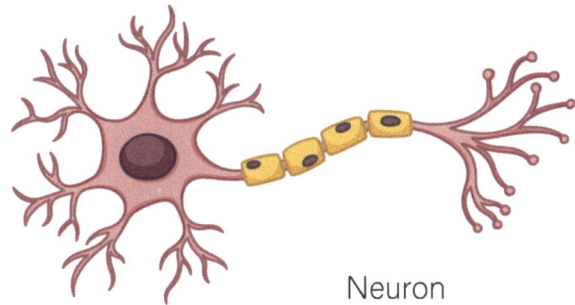
Neuron

One type of glia are star-shaped and secrete signals that convey and control how synapses form, function, and fade away.

Another type produce the insulating sheath of the axon and increase the rate at which electrical impulses pass on.

Neurons have a long cable, several times thinner than a human hair, called axons.
Synapse is the point of connection and communication between two neurons.

The first transgendered honoree of the National Academy of Sciences, Barres tells
of his career as devoted to solving "the mystery and magic of " glial cells.

He spoke for underrepresented women and minorities in STEM and wanted the system fixed.
"Science proceeds at its best when it includes diverse studies performed by diverse scientists", he said.

Defining Modern Computing

Alan Turing

Alan Turing, born in London in 1912, stayed away from parents and grew up with a retired army couple in England.

Alan's father worked in the British Indian Civil Services. His parents traveled frequently to India for such purposes.

He developed a love for mathematics and science early on. Could solve advanced problems; recognized as a genius thereupon.

He graduated in mathematics from King's College in Cambridge. Then, obtained a PhD from Princeton University in his quest for knowledge.

Alan introduced Turing machines for research into math foundations. Now deemed the original idealized model of a computer by historians.

His abstract computing device handles symbols on a strip of tape according to a set of rules.

The machine's memory tape is assumed to be infinite. Divided into cells each holding one symbol from a finite set.

A Bombe

Turing Machine

The machine's "head" would read from a cell
and act per its state as in "if in state A, and read X, then do L".

During World War II, Turing was part of the England's team of codebreakers
who intercepted the secret radio communications of the Axis Powers.

Alan specified an electromechanical machine that was automated
called the bombe, it could decrypt the messages secretly coded.

By the end of the war, more than two hundred bombes were in operation.
Conflict shortened by over two years, thanks to the acquired information.

Alan defined a standard for an intelligent machine with the Turing Test.
Alas in 1952, he was persecuted for his sexuality and threatened with arrest.

Alan developed many theoretical computing devices in spite of social impediments,
left an enduring legacy in the field of computing and artificial intelligence.

Researching AIDS Prevention

Bruce Voeller

Biochemist and gay-rights activist Bruce Voeller
was an AIDS research pioneer.

He was born in Minneapolis and raised in Oregon.
An accomplished pianist, skier, swimmer, and horseman.

He earned BS from Reed College and PhD in biology from Rockefeller.
Then continued his research at the university as an associate
professor.

He co-founded the National Gay Task Force, taking his career to new
heights.
He served as its executive director until 1978 campaigning for gay
rights.

Then, he co-founded the Mariposa Education and Research Foundation.
As AIDS flared, Voeller started research on preventing its transmission.

In 1981, the global AIDS pandemic began.
It remains an ongoing worldwide public health concern.

Within the first few years of the pandemic, Voeller coined the name
Acquired Immuno-Deficiency Syndrome (AIDS); with use still the same.

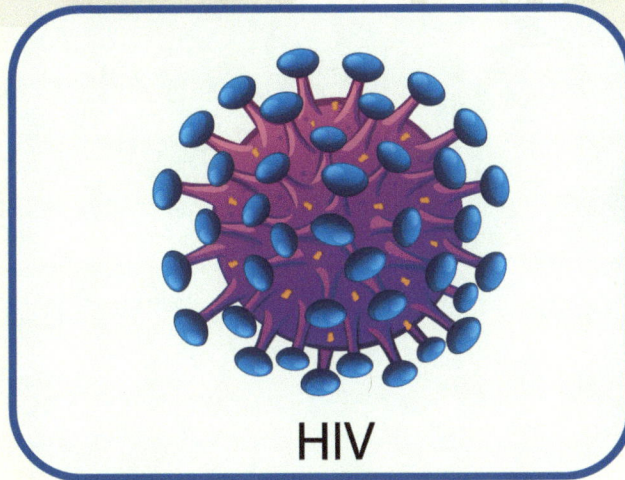

HIV

The human immunodeficiency virus (HIV) causes AIDS.
It infects the immune system and destroys T-cells as it invades.

Controlling and shaping the immune response, T-cells
are an important type of white blood cells.

AIDS spreads during pregnancy and birth from infected mother to baby.
Or, by contact with infected blood or some bodily fluids maybe.

AIDS shows early flu-like illness followed by progressive failure of the immune system.
Life-threatening opportunistic infections and cancers then thrive within the victim.

At Hunter College and Cornell University, Voeller pioneered research and testing
for effectiveness of protective barriers in preventing AIDS from spreading.

Voeller died of an AIDS-related illness... making us ponder.
In 2019, he was memorialized on the US National LGBTQ Wall of Honor.

Enabling Digital
Governance
Audrey Tang

Audrey Tang is an open-source programmer and a cabinet minister in Taiwan
dedicated to transparent democratic governance for citizens to rely on.

Before the age of five, Tang could read books of literary classics.
By six, she was on to advanced mathematics!

As a child prodigy, before age eight, she started programming computers.
She learned Perl, a general purpose open source language, with no tutors.

She is self-educated and in her teens held positions with ease
in California's Silicon Valley in her own startups and in other companies.

Tang initiated and led a project to implement the Perl 6 language, now called Raku.
She adapted many free software programs to different languages and regional needs too.

Living on the "net," she had been avoiding a conflict between societal norms and herself.
Finally, she transitioned to a female to reconcile her appearance with her inner self.

In 2014, to support a protest movement, Tang hacked a solution.
She broadcast protesters' messages across Taiwan with her contribution.

17

Instead of treating this hacking episode as a menace,
the Taiwanese Government hired her for her excellence.

Tang was invited to build digital media competence curricula.
It was implemented in 2017 in schools as a new success formula.

With the role to bridge the gap between the two generations, older and younger,
Tang was appointed as the youngest minister without a portfolio...
What a wonder!

Her mission is transparency and open government as the digital minister.
She is the country's first transgender and non-binary cabinet member.

Her initiative, *vTaiwan*, enabled use of social media paradigms by citizens.
Giving them access to influence regulations through digital petitions.

She tackled disinformation during the Covid pandemic with a public dashboard.
Her *Humor over Rumor* initiative used memes and fun to get facts explored.

Proponent of broadband as a human right, Tang ensures 10MG uplink and downlink
for democracy needs symmetric information with citizens uploading fact checks to think.

Bridging Health Disparities

Valerie Stone

Valerie Stone is a professor of medicine, an American physician, and a nationally recognized expert on HIV/AIDS condition.

As a teen she lost her grandmother to stage 4 cancer.
Motivated, Stone became focused on medicine as a career.

She studied at the School of Medicine at Yale University,
And completed residency at Case Western Reserve University.

At Boston City Hospital, she did fellowship in infectious diseases.
There, as her first job, she directed ambulatory care or outpatient services.

She saw family members, friends and classmates get HIV infection.
That drew her to AIDS care to save the community from devastation.

She treated her first HIV patient in 1983 when limited treatment pathways existed.
In 1996, highly active antiretroviral therapy (HAART) was invented.

HAART uses multiple drugs as a strategy to control HIV disease.
It enabled AIDS to be managed like another chronic illness such as Diabetes.

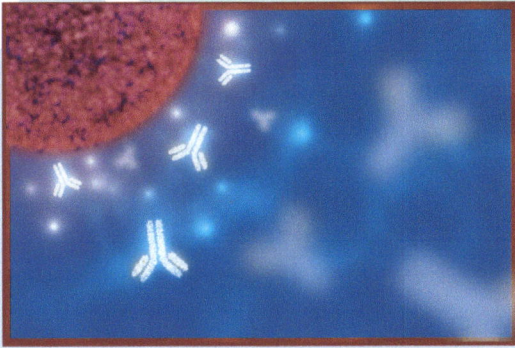

As an appointed faculty at Harvard University,
she researched the prevalence of HIV/AIDS in Black community.

Her focus was on disparities in care by gender, race and ethnicities
And how to optimize the care for patients from underserved
communities.

With her exceptional contributions in primary care research and
education,
she became Harvard's first African-American Professor of Medicine in
2011.

She worked as a senior scientist and a primary care physician
at the Massachusetts General Hospital, in addition.

Since 2019, she serves at the Brigham and Women's Hospital
as vice chair of diversity, equity and inclusion supporting human capital.

She shattered many glass ceilings for queer women and women of color.
She authored the book, HIV/AIDS in U.S. Communities of Color.

She was honored by the Massachusetts Medical Society
for the 2020 LGBTQ Health Award for fostering equality.

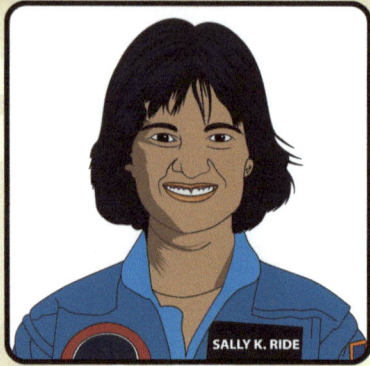

Venturing Into Space
Sally Ride

Sally Ride, an astrophysicist inducted into the Astronaut Hall of Fame,
has the title of the first American woman in space to her claim.

Her father, a World War II veteran, was a political science professor.
Her mother worked at a women's correctional facility as a counselor.

Ride initially aimed to become a professional tennis player.
Earned an Intercollegiate Women's Singles champion in her freshman
year.

While she didn't continue with professional tennis,
her foray into the sport gave her good physical fitness.

At University of California, she was the only woman majoring in Physics.
She transferred to Stanford as a junior and graduated since.

Doctorate in application of physics to astronomical observations,
she was a worthy applicant when NASA changed its selection
regulations.

Ride was selected out of thousands of applicants in 1978
to become part of the 35 selected for NASA astronaut group 8.

She was one of the only 6 women chosen to train for space-flight.
Ride's scientific expertise made reaching space within her sight.

21

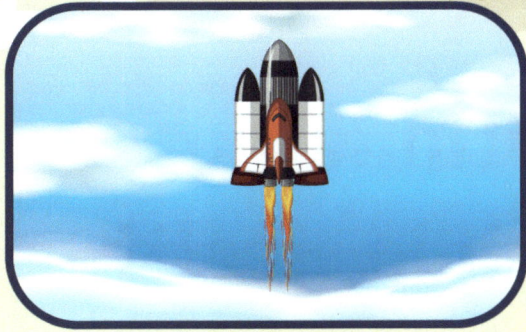

Ride served as the ground-based Capsule Communicator,
liaising between the astronauts in space and the Mission Control Center.

She helped develop a robot arm called the Shuttle Remote Manipulator System.
Assigned to her first space mission to work the robotic arm with her wisdom.

Aboard the Space Shuttle Challenger in 1983, Ride was the youngest astronaut ever.
She maneuvered the robotic arm to put satellites into space... what an endeavor!

After NASA, Ride helped women and girls achieve their careers in STEM.
She taught at University of California, and ideated NASA's EarthKAM system.

EarthKAM enhanced middle school students' education,
by allowing pictures of Earth taken using a camera on the International Space Station.

Sally Ride Science set up with her partner, Tam O'Shaughnessy, trained teachers.
She received, on behalf of Ride, the Presidential Medal of Freedom meant for achievers.

Optimizing Public Health
Rachel Levine

An admiral in the U.S. Public Health Service Commissioned Corps, Rachel Levine is
the first openly transgender four-star officer in any United States uniformed services.

At her all-boys prep school, she played hockey and did dramatics.
At Tulane University, she discovered her passion for pediatrics.

Pediatrics is a branch of medicine dealing with children and their disease.
Finding teenagers challenging and stimulating, she focused on them with ease.

After completing residency and fellowship at New York's Mount Sinai Hospital,
she held staff roles and did private practice to fulfill her goal.

Focusing on the intersection between mental and physical health,
as a physician, she treated children, adolescents, and young adults.

At Penn State, specializing in the treatment of eating disorders,
she created the Hershey Eating Disorders Program to address unhealthy behaviors.

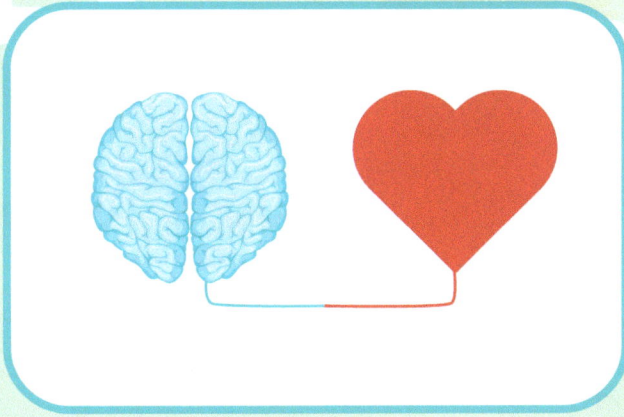

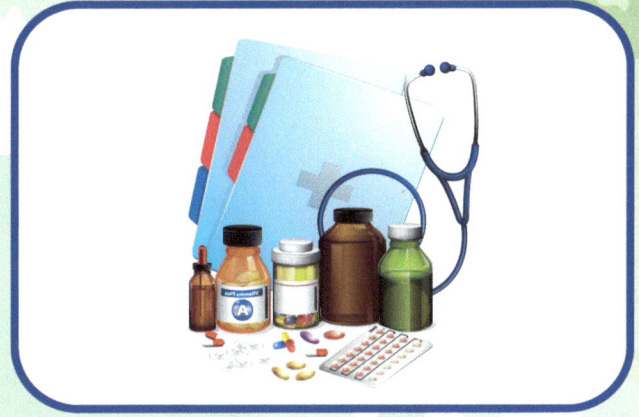

Preoccupied and anxious about food and how their body is looking, victims show restrictive eating, binge eating or purging by vomiting.

She also served as the Professor of Pediatrics and Psychiatry at the Penn State College of Medicine gaining notoriety.

Appointed as the physician general, she became the operational head and the senior spokesperson on Pennsylvania state's public health.

Under Levine's leadership, law enforcement officers carried life-saving medication to reverse overdose from substances like opioid.

Later, as the state secretary of health, Levine steadfastly led Pennsylvania's public health response on the global pandemic, Covid.

As the 17th assistant secretary for health in the US government administration,
Levine is the first transgender officer to receive US Senate confirmation.

Levine is addressing health disparities among LGBTQ youth including bullying, suicide, discrimination, isolation and medical distrust.

THE END

www.ingramcontent.com/pod-product-compliance
Lightning Source LLC
Chambersburg PA
CBHW041556040426
42447CB00002B/194